NUMBER 224

THE ENGLISH EXPERIENCE

ITS RECORD IN EARLY PRINTED BOOKS
PUBLISHED IN FACSIMILE

WILLIAM BEDWELL

MESOLABIUM ARCHITECTONICUM

LONDON 1631

DA CAPO PRESS
THEATRVM ORBIS TERRARVM LTD.
AMSTERDAM 1970 NEW YORK

The publishers acknowledge their gratitude
to the Governors of the John Rylands Library
Manchester, M3 3EH
for their permission to reproduce
the Library's copy
(Shelfmark: 6909.7)

S.T.C. No. 1796
Collation: A2, B4, C2, ¶4

Published in 1970 by
Theatrum Orbis Terrarum Ltd.,
O.Z. Voorburgwal 85, Amsterdam
&
Da Capo Press
- a division of Plenum Publishing Corporation -
227 West 17th Street, New York, 10011
Printed in the Netherlands
ISBN 90 221 0224 6

MESOLABIVM ARCHITECTONICVM

THAT IS,

A most rare, and singular Instrument, for the easie, speedy, and most certaine measuring of Plaines and Solids by the foote:

Necessary to be knowne of all men whatsoeuer, who would not in this case be notably defrauded:

Inuented long since by *Mr. Thomas Bedwell* Esquire:

And now published, and the Vse thereof declared by *Wilhelm Bedwell*, his nephew, *Vicar of Tottenham.*

LONDON,

Printed by *I. N.* for *William Garet*

1631.

TO THE ILLVSTRIOVS,

Right-honourable, Right-worshipfull, and dearely beloued, the Nobility, Gentry, and Commons of *Great Britaine*, and *Ireland*.

GOd, sayth the wise man, hath ordered all things by measure, number, and weight. And man, the image of God, ought, as the Philosophers teach, to order all his life according to the same directions. And yet who knowth not, how little they are of all men regarded! To passe by the generall, and to come to that which concerneth our commerce, What smatterer in the Mathematicks is hee, who knoweth not, what neglect or ignorance there is, euen in those artists, whom all men, the Rich aswell as the Poore, do, and must daily trust, in matters of measuring! I accuse no man of wilfull fraude or malice. But this I say, There is no man whatsoeuer, that is not some peece of a scholler, that can measure tymber truely: And those who are most skil-

full in both, cannot do it either speedily, or readily. All which, Illustrious, Righthonourable, Right-worshipfull, and Dearely beloued, I promise in this short treatise, by the ordinary Instrument, in this case vsed, to teach the meanest of vnderstanding, though wholy vnlearned, to do, with that speede, facillity, and certainety, that may not be bettered. This as a prodromus, *begun and ended, in the middest of many and great troubles, I thought good to premise and send out, before a larger discourse of the Fabricke, and more ample Vse thereof, which, God willing, shall follow, so soone as Figures and Diagrammes may conueniently be cut, for that purpose, with all possible speed: In the meane time the Author, wholy deuoted to his Countries seruice, resteth*

Your H.H.H. in all obseruancy,

Wilhelm Bedwell.

MESOLABIUM ARCHITECTONICUM.

CHAP. I.

Of the Mesolabe: And of the vse of it in generall.

1 To measure by this Rule, is by two knowne lines, to finde out the third vnknowne.

THe Instrument whose vse at this time wee intend to declare, is no other, in respect of matter and forme, in generall, but the Carpentars rule, by them vsed in the measuring of Tymber, and Bourd by the Foot square: For it is a flat Ruler, or oblong parallelogram, of two foote, or a foot & halfe long: Two inches and an halfe, or there abouts, broad: And of such convenient thicknesse as shall at euery mans discresion be thought most fit.

Againe, as theirs, so this on the one side, contayneth a Scale of equall diuisions, First of Ynches, Halfe-ynches, Quarters, Halfe quarters, and so forth: Then againe, on the same side, you haue an Ynche diuided into Seauen, Eleuen, Thirteene, Seauenteene, Nineteene, and Three and twenty, and such other equall parts, as euery man for his owne vse shall think most fitte, and the workmans hand shall be able to performe.

More ouer, on the other side, as on theirs also, you haue a Scale of vnequall diuisions, seruing for the measuring of Bourd and Tymber: But after a farre different manner: For their diuisions are only markes or small strokes, in one of the limbs of that side, determinyng from the Fore-end of the Rular in ynches, and partes of ynches, the Square measure of solids or Tymber. Wheras this of ours consisteth of two sortes of straight lines, the one Beuelling or Slanting, drawne askue from side to side: The other Parallell that is equidistant one from another running along the Rular, from the one end toward the other: And therefore cutting those former, and diuiding them into vnequall portions, where by not onely their sayd Quadrate or square measure is performed: But also all other whatsoeuer, and that with great facillity, speede, and certainety.

Lastly here, as also there, you must make a distinction betweene end, and end; For that end we call the For-end of the Rular, from whence the diuisions of it into ynches, on both sides are begun to be reckoned: And that the Backer-end where they doe end and determine: Or, contrarywise, the For-end is that from whence the numbers ascribed to the Beuelling linnes are lesse and lesse. But the distances betweene them are greater and greater.

Thus much of the Ruler, and the Partes therof. *Mensura, innuit Aristoteles, in quolibet mensurabili genere, est quippiam minimum*: A measure, as Aristotle seemeth to intimate, is some small portion in euery thing that is to be measured: And it is commonly termed of the Geometricians *Famosa mensura*: A knowne, or set measure generally agreed vpon amongst all men: As in measuring by hand-breadths, feete, and passes, one hand breadth, on foot, one passe. And in deed it is an old saying of *Protagoras*, as Aristotle recordeth, *That man is the measure of all things*. And true it is, That Vitruuius, and Hero the mechanicke or inginer, do shew, That generally all measures are taken from the partes of Mans body, as a Finger, an Ynch (*Pollex*) an Hand, or Hands

breadth

breadth, a Spanne, a Foot, a Cubite, a Passe, an Elne, a Fathome.

But who knoweth not, What great difference there is between man & man? And not only between men of diuerse Countreys and climats: But eu'n between those of one and the same prouince; Nay of one and the same family, children of the same parents? And, the limmes of men being proportionall to their bodys, what difference must there needs bee, betweene the measures taken from them? And in deed heerupon it came to passe, That the Measures, not only of diuerse Nations: But eu'n of one and the same, are, and alwayes haue beene much different, as doth manifestly appeare by the diligent comparisons made of them by diuerse and sundry learned men, and especially by that hopefull Willebrordus Snellius, as wee shall, God willing, shortly teach in Ramus's Geometry, which wee purpose to set out in English, for the benefite of such of our Countrey men, as delight in these study's, yet are ignorant of those languages where in they are written.

This difference was in this our kingdome complained of in all ages: For from hence arose many greeuous quarrells and suites in the Law, which our worthy Kings, and state in their Parlaments, in all ages haue laboured to appease, by reducing all to an vniformity: For thus wee finde in our Statutes: *It is ordeined, That 3 grains of* Barley, *dry and round, do make* an Ynche: *Twelue ynches do make* a Foot: *Three foote do make* a Yard: *Fiue yards and halfe do make* a Perch. *And 40 perches in length, and 4 in Breadth do make* an Aker. 33 of Edward the first, *De Terris mensurandis* Item, *De Compositione vlnarum et Perticarum.* Againe in a Perlament held in the 25 th of Queene Elizabeth, you haue an Act, thus intituled: An Act for the restrainte of New-buildings, &c. in & nere the citys of London & Westminster *Be it enacted by the authority afor'said,* That a Mile *shall be taken & reckoned in this manner, & no otherwise: That is to say,* a Mile *to containe 8 Furlongs. And euery* Furlong *to containe 40 lugges or poales: And euery* Lugge *or Poale, to containe 16 foot and an halfe.*

 Al-

Although this same our Rule may bee fitted for sundry other sortes of measures: Yet we haue here nothing to do, But with the Foote, and his partes, which are Ynches, Halfe-ynches, Quarters, Half-quarters, and such other sensible partes of the same.

2 Things to bee measured by this Rule, are magnitudes.

3 A magnitude is a continuall quantity.

A magnitude, or a bignesse is that which hath one, or more dimensions: Now dimensions are in number three, to weet Length, Breadth, and Thicknesse.

4 A magnitude is of one dimension, or many.

5 The measure is of the same nature with the thing to be measured.

6 A magnitude of one dimension is called a Line.

A line, is a magnitude of length onely. Or, A line is a magnitude onely long. Such are wayes, or distances betweene place and place. Such a magnitude, sayth Proclus out of Apollonius, is conceiued in the measuring of iourneys. And by the difference of a lightsome place, from a darksome. Such are Lenghts, Heighths, Depths, and Breadths. Therfore here

7 The measure vsed is a line.

Here therefore there is no further skill required in the measurer then a due application of the measure giuen: And therefore here in this case there is not any vse of this our Instrument.

CHAP. II.

Of the measuring of Plaines by the foot square.

1 A magnitude of many dimensions, is of two or three: That is called a Surface: This a Solid.

2 If a dimension giuen, be eyther greater, or lesser, then any of the numbers vpon the Rular, you must take some lesser, or greater,

which

which is proportionall vnto it.

3 A surface is a magnitude long and broad.

That is, a surface is a magnitude which hath two dimensions, to weet Length and Breadth. Such magnitudes, sayth Apollonius, are the shadowes vpon the ground, which ouerspread the fields farre and wide, but do not enter into, or pierce the earth: Neither haue they any thicknes at all. The Greek woord *Epiphania*, is here more significant. For this worde intimateth no more but, The outward appearance of any thing. For of a magnitude nothing is to be seene but the surface. Such are bourds esteemed to be by the Carpentars: Wainscotte, by the Ioyners: Glasse, by the Glasiers: Cloth, both linnen & Woollen, by the Drapers: Land, Medowe, & Wood, by the Surueighers: For in the measuring of these, there is only Breadth & Length considered, with out any respect at all had to the Thicknesse. Therfore

2 Here the measure is a Surface.

Surfaces, according to their diuerse natures, are measured with diuerse and sundry kindes of measures: Wood, Land, & Medowe, are measured by the Rod or Perch: Cloth, Painting, Pauing, & Wainscotte, by the Yard: Bourd and Stone, by the Foote. Although this our Instrument may be fitted to all these, or any other like measure, Yet wee at this time intend to meddle with no other but the last, to weet With the Footesquare.

4 A surface is either Plaine or Vneu'n.

5 A Plaine surface is a surface, which lyeth equally between his bounds.

A surface, the learned knowe is geometrically made of Lines: Therfore as lines are either straight or Crooked: So from hence are all surfaces Straight or Crooked: Or, to speak more properly, Eu'n or vneu'n, Plaine or Rugged: Yea & by a straight line are surfaces tried, whether they be Eu'n, or vneu'n. For if a rightline applyed to a surface euery way, do touch it in all places, it is Eu'n: Otherwise, it is vneu'n.

 Plaines,

9 Plaines, as wee sayd, are measured by the Foote square, That is the quadrate of 12 ynches.

A foote of plaine or flatte measure is the quadrate of 12 ynches, or that which is equall vnto it. That is, it containeth 144 square Ynches: For 12 times 12, are 144. Hauing therefore a plaine giuen of 12 ynches broad, there is no question but 12 ynches of that breadth shall make a Foote. But if the breadth giuen be greater or lesse then 12, there is a question, What length, with the breadth giuen, shall make a plaine equall to the square 144. Here

7 Of the two lines giuen, the one is the breadth assigned, the other is alwayes the beuelling line 12.

Here againe it must bee remembred, *That onely those plaines are to be measured which are Rightangled parallelogramms*, Or to speake in their owne Language, which are comprehended of a, Base, and Heigh which are rationall betweene themselues: Ramus 9 e IIII. Those plains therfore which are not such, must bee reduced vnto these kinde of figures.

I An example or two shall make all plaine. A bourd of 16 ynches broad and 18 ynches long, (And so a stocke of 13 bourds) is to be measured. Here I finde 16, the line answering to the Bredth, to crosse the beueller 12, at 9 ynches from the fore-end of the Rular. Therefore I say euery 9 ynches of that length shall make a Foot of bourd: Or which is all one, shall be equall to 144, the square of 12 ynches. Now 9 ynches I finde to bee contained in 18 foote, the Length, 24 times: Therefore I say, The bourd assigned doth containe 24 foote of bourd. Lastly, there being in the stocke 13 such bourds, I say the whole stocke doth contayne 312 foot of bourd.

II A Table of 36 ynches broad, and 28 foote long, is to be measured. Here 36 is greater then any of the parallels found vpon the Rular: Therefore by the 2 e of this, I take

take 18 the halfe of it, which I finde to meete with 12, the beuealing line, at 8 ynches from the for'end of the Rular: Therefore euery 8 ynches of length, of the bredth 18, shall contayne a foote of bourd: But the breadth giuen is 36 ynches: That is twice 18: Therefore euery 8 ynches in length, of that Table shall be 2 foote of bourd. Now againe I finde 8 ynches, in 28 foote 42 times: Therefore the Table containeth twice so many foot: That is 84 foote of bourd.

III A pane of Glasse, 7 ynches broad, is to bee measured. Here 7 is lesser then any of the parallels: Therefore by the 2 e of this, I take 14, the double thereof: Which I obserue to meete with 12, at 10 ynches and 2 seauenth parts of an ynch from the fore-end: Therefore euery 10 ynches and 2 seauenth partes of an ynch, of 14 ynches breadth, shall bee a foote of Glasse: But the breadth giuen is but 7 ynches: Therefore euery 10 ynches, and 2 seauenth partes of an ynch shall be but halfe a foote of glasse.

Of the measuring of Triangles, and all other Rightlined plaines.

8 A triangle is nothing else but the halfe of a quadrangle, or parallelogramme: And if it haue one right angle, it is the halfe of a rightangled parallelgramme. *Therefore*
9 It is to bee measured as the Rightangled-parallelogramme, onely conceiue that the number found, shall bee the double of that which is sought.

Here therefore it must bee conceiued, That of the two sides encluding the Rightangle, the one is to be vnderstood to be the Breadth, the other the Length.

I Suppose a Rightangled-triangle, whose sides including the Right-angle, are 18, and 24, are to bee measured. Here I take 18 for the Heighth, or Breadth of the parallelogramme

gramme, which also I finde to meete with the beuelling line 12, precisly at 6 ynches from the fore end of the Ruler: Againe 6, the sayd line found, I finde iust 4 times in 24 the Lenghth giuen: Therfore I auerre the Triangle giuen to conteine the halfe of 4 foote, that is 2 foote of bourd.

20 If the triangle giuen bee not right-angled, then is it by a perpendicular, let fall within the triangle, from one of the corners vnto the base, to bee reduced vnto two rightangled triangles.

How this is to be done, *Euclide* teacheth at the 11 & 12 propositions of his I, booke; And P. Ramus, at the 9 & 10 elements of his V. booke of Geometry. It is also to bee done by the squire. Or by a triangled leuell, and otherwise.

II An Obtusangled triangle, whose three sides are 25 40, and 42, is to bee measured. Heere by one of those aboue named wayes, I finde the perpendicular or plumbline, falling from the greater corner, vnto the opposite line, to be 24. And 24 I finde vpon the Ruler to meete with the line of 12, at 6 ynches from the fore-end of the same: Againe 6 I find in 42 seauen times: Therefore the Triangle giuen doth conteine halfe so many foote, That is 3 foote and an halfe of bourd.

11 From hence it is manifest how any Rhombus, Rhomboides, Trapezium, or irregular rightlined multangles are to bee measured.

To weet, that they are to be measured by parts, or by the particular triangles, which euery such figure doth contayne. Examples you may haue in the XIIII booke of Ramus's Geometry, or in any others, which haue written of Geometry.

Of the measuring of any ordinate multangle figured.

12 Ordinate multangled plaines are measured by

by their halfe Perimeter, and the plumbline from the center, vnto the middest of any one side.

These sortes of plaines may bee measured, as the former were, by diuiding them into their seuerall Triangles: But this last is farre shorter: And therefore to bee embraced & rather to be vsed in practise. Here the halfe of the perimeter, or bout-line, answereth to the Length in a parallelogramme: And the plumbline here, is in stead of the Heighth or Breadth there.

I An ordinate Pentangle, whose sides are 24 ynches a piece; And the Plumbline from the center, to the middest of any one of the sides 16, is to be measured. Here 16 the Plumbline or Heighth, doth, vpon the Rular, meet with the slanting line 12, at 9 ynches from the oft named end: And 9 is contayned in 60, the halfe of the perimeter, 6 times and two thirds: Therefore the Pentangle giuen conteineth 6 foot, and two third partes of a foot of Bourd. II A Sexangled ordinate figure, whose sides are 12 ynches broad a piece, is to bee measured. Here the Plumbline from the center to the middest of any one side, is 10 ynches, and 8 one and twentyths of an ynch: The double of 10 (that is 20.) and 16 one & twenty parts of one ynche, I obserue to meete with the beueller 12, about 7 ynches, & one quarter of an ynch, from the fore end of the Rular. Which 7 and a quarter, is contained in 44 six time, and two twenty nineth partes. Therfore I say the Sexangled figure giuen doth containe 6 foote of bourd, and some small quantity more. The Circle, or Circular forme is in like manner measured: For

13 The Circle is measured by the Ray, and the halfe of the perimeter.

For, sayth the Geometrician; *Planus e radio & peripheria dimidio est area circuli*. The plaine of the ray, and halfe of the circumference is the content of the circle. A Round table, whose diameter is 4 foote, and 8 ynches, (or 56 ynches) is to

 bee

be measured. The halfe of 59 is 28: And the halfe of the circumference is 88. Now 28 being geater then any of the paralells, I take 14 the half therof: Which I find to meet with the beuelling line 12, at 10 ynches, and a quarter, from the for'end of the Rular: Therfore I say euery 10 ynches, and a quarter of an ynche of that Table shall be 2 foot of bourd. And because 88 doth containe 10 and 1 quarter, 8 times, and 20 fourty ones; Therefore I say, the whole doth containe 16 foot of of bourd, and 144 ynches.

CHAP. III.

Of the measuring of Bodies or Solids by the Foot.

1 A Body is a magnitude of three dimensions. A Body or Solid is a magnitude which hath Length, Breadth, and Thicknes.

2 Here the measure is also a body, to weet the Cube of 12 that is 1728.

This is our opinion: Yet if any shall thinke it a paradox, or shall gaine say it, or mainetaine the contrary, wee will not contend. And

3 Of the three dimensions, two are giuen, the third is sought.

4 Bodies are of diverse sorts: But we will at this time meddle only with such as are comprehended of parallelogrammes, or with Cylinders.

True it is, that this our instrument may bee fitted, and applyed to the measuring of many other sorts of Solid bodies: But because we see no great vse of it in the measuring of any other, then of these two sorts: Therefore wee will declare the vse of it, in the measuring of these two onely. Of these the first is the Parallelepipedum, which is a plaine Solid, whose opposite sides are parallelogramme.

I A rightangled parallelepipedum (or a squared tymber logge) of 12 Ynches thicke, 18 broad, and 16 foote long, is to be measured. Here the Thicknesse and Breadth are giuen: The Length is sought. These I finde vpon the Rular to meet at 8 ynches from the oft named fore-end: Therfore I Say, Euery 8 ynches of that Logge in length shall make a solid foote of tymber. And because I finde 8 Ynches, in 16 foote, 24 times: Therfore I say in the Tymbersticke giuen, there is 24 foote of solid measure.

II A squared stone of 14 Ynches thicke, fiue foote (or 60 ynches) broad, and 10 foote long, is to be measured. Here 60 is greater then any of the parallels vpon the Rular: Therfore I take 12 the 5th part of it: And I obserue 12 and 14, to meete at 10 ynches, and 2 seaunth partes of an ynche, from the Fore-end of the Rular. Therfore I say, That euery 10 ynches, and 2 seaunth partes of an ynch in length of that stone shall be 5 foote of solid measure. And because that 10 foote conteineth 10 ynches, and 2 seaunth parts of an ynche, 11 times and 5 seau'nty twoos: Therfore I say the whole stone conteineth 58 foote, and one third parte of a Foote of solid measure.

III A rightangled Prisma, both whose sids, Parallelogramm's I meane, conteyning the rightangle, are 18 ynches broad; the whole being in length 16 foot, is to be measured. Here vnderstand that, as before was shewed, as a Triangle was but the halfe of a quadrangle: So a Prisma is nought but the halfe of a Parallelepipedum, sawne longways from corner to corner though the midd'st: And hence in Greek it hath the name: This knowne I enter with the numbers giu'n, and I finde 18 to meet with 18, at 5 ynches and one third parte of an ynche from the oft named end of the Rular: Therefore I say, That euery 5 ynches, and 1 third parte of an ynche in length of that sticke shall be but halfe a foote of solid measure. Nowe because 5 ynches and 1 third of an ynche is conteined in 16 foote, 67 tymes and 14 sixteen partes, that is almost 68 times: The fore I say, The Prisma giu'n doth conteine almost 68 halfe foot's, or 34 foote of solid measure.

IIII A sispaned solid, all whole sides are 6 ynches broad a peece and 16 foote long, is to bee measured. Here the two lines giuen are, as aboue was taught, the Plumbline from the center, vnto the middest of any one of the sides: And the halfe of the compasse; That, as before was taught, is 5 ynches, and 2 eleuenth partes of an ynche: This is, as you see 18. Now 5 and 2 eleu'nths doth meet with 18, at 19 ynches and 1 fifth parte of an inche from the fore-end: Therfore I say, That euery 19 ynches, and one fifth parte of an ynch, shall be a foote of solid measure. Lastly, because 16 ynches, and 1 fifth parte is conteined in 16 foot, 10 times, and 2 fifteene pates, I say that the tymber sticke giu'n doth containe 10 foot of solid measure, and some small quantity more.

Lastly a Round columne, or Cylinder, of 44 ynches about, & 12 foote long, is to be measured. Here according to that aboue taught, the two lines giu'n are, The half diameter, & the halfe circumference: This is 22: That 7. Now these two do meete vpon the Rular at 11 ynches, and 17 seauenty two partes, of an ynch, from the fore-end there of; Therefore the sticke containeth about 13 foot of tymber or solid measure.

***

FINIS.

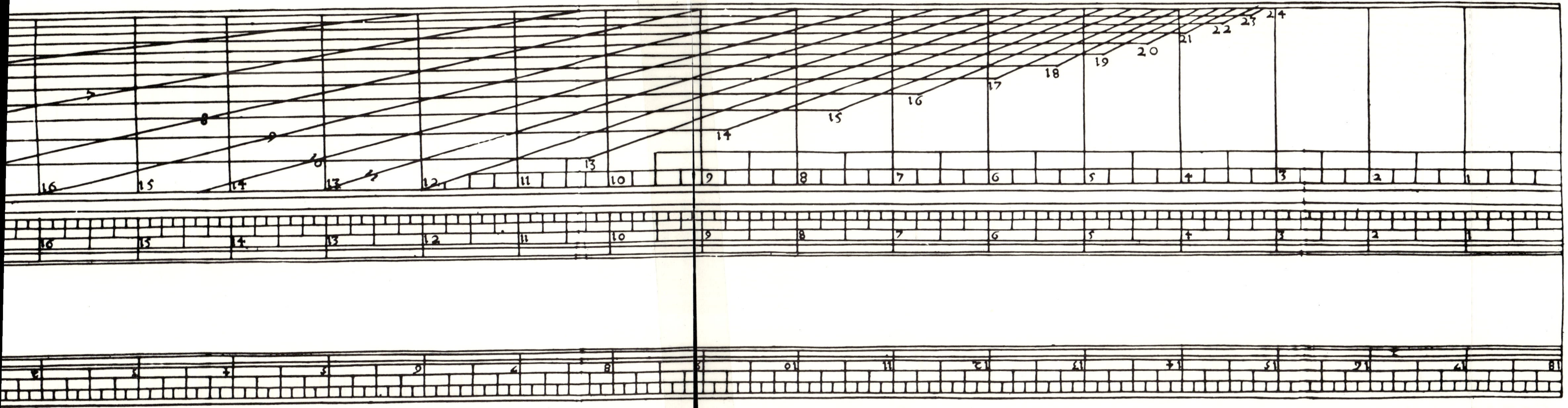

AN
APPENDIX
TO THE
MESOLABIVM.

TWo things, for the further illustration of the Instrument, wee haue thought good here to annex vnto the former. The one is a collation of this manner of measuring, with that commonly taught and practised. The other is of the Measuring of Land by the Aker.

A foot of solid measure, as all doe generally know, is the Cube of 12 ynches; that is, a square, or dye-form'd body, all whose dimensions, to wit, Thicknesse, Breadth, and Length are equall: And the content, in numbers, is found by a continuall multiplication of 12, 12 and 12, thus: 12 times 12, are 144: and 12 times 144, are 1728. Therefore a foot of solid measure doth conteine 1728 seuerall cubes of an ynch, thicknesse, breadth, and height. This is as it were the standard, whereby this kinde of measuring is to be examined.

All artificers generally do measure by a Table of square numbers. And therefore if the body giuen to be measured, be square, that is, if the thicknesse and breadth bee equall, they can giue the iust content. But where these two dimensions doe differ, there by their rules they doe it not without some error. For to bring it to the vse of their Table, they must first make the thicknesse and breadth equall; which they doe, by taking the excesse from the greater, and adding of it to the lesser, or, which is all one, by girding of the body about, and by taking of the quarter of the compasse. An example or two will make all plaine.

Suppose a body, giuen to bee measured, were 10 ynches, thick, and 14 broad. Here they take 2 from 14, and adde it vnto 10, and so doe make all the sides equall: Or by girding of it they finde the compasse to be 48. And the quarter of 48 to be 12. And their table for the square of 12 ynches, doth giue 12 ynches for the length required to make a foot of solid measure. If this be true, then 10, 14, and 12, continually multiplyed between themselues, shall be equal to 12, 12, & 12, continually multiplyed between themselues. But 10, 14, and 12, doe giue for the product 1680. And 12, 12, and 12, as in former wee saw, gaue 1728. And the difference betweene 1728, and 1680, is 48. Therefore the losse in euery foot of that body, by their measure, is 48 inches.

Here, vpon our Rular, you see 14 the parallell, to meet with 10 the beuelling line at 12 inches, and $\frac{12}{35}$ of an inch from the fore end of the Rular: And by multiplication you shall finde that 10, 14, and 12 $\frac{12}{35}$ doe make 1728: Therefore our rule is true.

If the body giuen be 8 inches thicke, and 16 broad, they likewise take 4 from 16, and doe adde it vnto 8, and so, as afore, doe suppose it to be equall to 12 inches square. Which, if it be true, then the product of 8, 16, and 12, shall bee equall to the product of 12, 12, and 12. But the product by the continuall multiplication of 8, 16, and 12, is but 1536: And the product of 12, 12, and 12, is 1728: and the difference betweene 1728, and 1536, is 192: Therefore by that their measure there is lost in euery foot 192 inches. Now vpon the Rular we finde 8 and 16 to crosse one another at 13 $\frac{1}{2}$: And

8,

8, 16, and 13 $\frac{1}{2}$ continually multiplied one by another, doe giue the product 1728: Therefore this kinde of measuring by the Rular is exact.

If it were 6 inches thick, & 18 inches broad, by the same reason, euery 12 inches in length should make a foot of solid measure. For the summe of all the sides added together, is 48: And the quarter of 48 is 12: And their Table for the square of 12, doth assigne 12 inches for the length. But 6, 18, and 12, multiplied continually doe make but 1296, which differeth from 1728 by 432. Therefore by this their measure, in euery foot of solid measure, there is lost 432 solid inches. That is iust one quarter of a foot.

Vpon this our instrument 6, and 18, are obserued to meet at 16 inches from the said fore end: and therefore it alloweth for a foot 16 in length. And 6, 18, and 16 continually multiplied doe make 1728: therefore our rule is true.

Of the measuring of Plaines, or Land, Meddows, and Woods by the Aker.

ALthough this instrument, as the title specifieth, bee fitted only for the measuring of Plains and Solids by the foot: Yet, as before is mentioned, it may easily bee applyed to other like sort of measuring. Now, among others, there being none of more frequent vse amongst vs, then the Rod or Perch for the measuring of Land, Meddow, and Wood by the Aker: And this being either not easie to be done by the vnlearned; or not speedily to be performed by any, I haue thought it not amisse, for the further declaratio of the vse & excellency of this inuention, and for the benefit of others, to adde vnto the former, something of this also.

An Aker of Land is, as before we heard, an oblong parallelogramme, whose breadth is 4 Poles, and length 40. Therefore an Aker conteineth 160 square Rods, of what figure or forme soeuer it be. For 4 times 40, are 160. This here in this case is, as 144 was in Board, and 1728 was in solid measure, as it were the Standard, whereby this kinde of measure is to be examined.

Land, Meddow, or Wood, is to bee measured by this instrument, in all respects, as Board or Glasse was measured Only two things are here first to be knowne: The first is, That as there the beuelling line of 12, was alwayes giuen for the breadth, as appropriate to that kinde of measure: So here another, peculiar to this manner of measure, is in like sort to bee drawne ouerthwart the parallell lines from $6\frac{2}{3}$ of an inch from the fore-end of the Rular, in the parallell 24, vnto 13 and $\frac{1}{3}$ in the parallell 12. The second is, That as there lines and spaces did answer, and were denominated of inches and parts of inches; so here the same lines and spaces must bee supposed to signifie Rods and Perches, and parts of the same. An example or two will make all plaine and manifest.

1 Suppose a right angled square field of 16 Pole broad, and 30 in length were to be measured: Here I finde 16, to meet with the line of Land measure at 10 inches from the fore end of the Rular: Therefore I say, that euery 10 rod in length, of the breadth of 16 rods, shall make an aker of Land. And againe, because the said 10 is found in 30 the length giuen 3 times: Therefore say, the field assigned doth conteine 3 Akers.

2 A square Meddow right-angled of 40 Poles in breadth, and 60 in length, is to be measured. Here 40, the lesser number of the two giuen, is greater then any vpon the Rular: Therefore I take 20, the halfe of 40: And I finde 20 to meet with the line of land measure, at 8 inches from the said fore-end of the Rular, Therefore first I say, That euery 8 Pole in length of the breadth of 40 Pole, shall conteine 2 Akers of

Meddow,

Meddow. Againe becauſe 8 is conteined in 60 ſeuen times and $\frac{1}{2}$: therefore I auerre, that the Meddow aſsigned to bee meaſured, doth conteine 15 akers.

3 Admit that a Wood to bee meaſured were 160 rods ſquare: that is, that euery ſide of the ſame were 160 Poles in length. Here 160 is farre greater then any number vpon the Rular: Therefore I take 16, the tenth part of 160: which I orſerue to croſſe the line of land meaſure at 10 Inches from the fore end of the Rular: wherefore firſt I affirme, that euery 10 pole of the breadth of 160 pole, ſhall conteine 10 Akers of Wood: Againe, becauſe 160 doth 10 ſixteene times, I ſay, that the Wood doth conteine 160 Akers. Or, which is all one, that euery Pole in breadth of that length, doth conteine an Aker.

But ſome man may obiect and ſay, this is not a matter worth the learning, or of ſo many words, ſeeing that it is well knowne, that there are many men wholly vnlearned, yea, and ſome of no extraordinary parts of capacity or vnderſtanding, which can meaſure Land, Meddow, or Woods, ſo that they be ſquare, or of any ordinary forme. I confeſſe I haue knowne diuers ſuch. And yet is not this our labour in vaine, or altogether vnprofitable: For firſt, what they doe with much ſtudy or contention of minde, and are long in doing of it, we teach to doe with great facility and ſpeed. For they, although the field be a rightangled parallelogramme, muſt firſt meaſure at the leaſt two ſides comprehending one or other of the rightangles. And then multiply theſe two ſides the one by the other: And laſtly, the product found, they muſt diuide by 160: And ſo by the quotient now found, anſwer the queſtion propounded. We ſaue a great deale of this labour. For we only meaſure the breadth of the field; then wee ſeeke vpon the Rular, how much in length of this breadth doth make an Aker. Laſtly, we apply this laſt number found vnto the whole length, and ſo without either multiplication or diuiſion, doe ſpeedily anſwere the demand.

Another thing there is wherein this inuention doth goe farre beyond the reach of the vnlearned, which is thus. These men can sooner measure and cast vp the whole content of the field, then they can set you out one, two, or three Akers of the same. Here it is all one to giue the content of the whole, or any part, or parts of the same. And that which to those is most hard, here to vs is most easie: I meane to set out any one or more Akers of the same; and that on which side or end of the field you shall thinke good.

Wilhelmi Bedwelli

Trigonum Architectonicum:

THE CARPENTARS RVLE,

Explaned, reformed, and enlarged:

That is,

A Table seruing for the more exact, and speedy measuring of Boord, Glasse, Stone, and such like, both Plaines and Solids, by the Foot, then euer heretofore hath in this kinde, beene set out or taught by any:

Inuented, and first published in the yeere 1612, by *Wilhelm Bedwell*, Parson of S. Ethelburghs London.

Of the vse of the Trigon in measuring of Plaines by the foot.

TO measure by this Table, is, by two numbers knowne, to finde the third vnknowne. Things here to be measured, are magnitudes. And magnitudes or bignesses, are quantities which haue either one or more dimensions, to wit, length, breadth, or thicknesse. These dimensions are here represented by numbers. *Of the two numbers assigned, the one must be sought amongst those on the out side of the Trigon: The other amongst those underneath the base of the same.* The numbers, as you see, doe stand against the spaces enclosed betweene two parallell lines. Hauing found your numbers giuen, let your eye direct you through the opposite spaces against which they stand, from the one ascending; from the other sidewaies, or contrarily, vntill you obserue both the spaces to meet, or fall into one. The number there found, is the dimension sought. *Plaines are magnitudes long and broad*: Or, Surfaces are magnitudes of two dimensions, namely, length and breadth. *Here the two numbers giuen, are that of the breadth, and the number of 12*; 1 A Board of 18 inches broad, and 24 foot long, is to be measured. Here the breadth 18 taken amongst those vnder the base, and 12 amongst those on the side, doe in the quadrate where their spaces meet, giue 8 inches for the length desired. Now because 8 inches is contained in 24 foot 36 times: therefore the board giuen, doth containe 36 foot of plaine measure. 2 A pane of Glasse is 8 inches broad. Here 8 taken amongst those on the side; and 12 vnderneath the base, doe set vs out 18 inches for the length. *If the breadth bee greater then 24, that is, then any number enclosing the Trigon, then take the halfe, one third part, one quarter, &c. and the number found shall be two, three, or foure foot, &c of plaine measure.* 3 A Table of one yard and a quarter (or 45 inches) broad is to be measured. Here 45 inches is greater then any number about the Trigon; therefore I take 15 the third part of the breadth, and 15 and 12 I finde to point to 9 and $\frac{3}{5}$ for the length desired. Therefore I auerre, that euery 9 inches, and $\frac{3}{5}$ parts of an inch in length of that table, shall conteine 3 foot of plaine measure. 4 A roome of 16 foot broad, and 48 foot long is to be floored; I would know how many foot of Board it will aske to couer it. Here 16 foot, that is, 196 inches is greater then any about the Trigon; therefore I take 16 the 12 part thereof: and 16 and 12 doe allow 9 inches for the length. Now because 9 inches are conteined in 48 foot 64 times; and 64 times 12 are 768. Therefore I say, the floore will require 768 foot of board to couer it.

The vse of the Trigon in the measuring of Solids by the foot.

SOlids or bodies haue three dimensions, to wit, length, breadth, and thicknesse. And of these commonly the breadth and thicknesse are giuen; the length is sought 1 A square timber stick of 12 inches broad, and 12 inches thick, is to bee measured. Here 12 and 12 doe point out 12 inches for the length desired. 2 A stone is 18 inches broad, and 16 thick. Here I finde 6 inches for the length sought. *If either one or both of numbers giuen, be greater then any about the Trigon, take as afore, either the one halfe, one third, and the number found shall answere in a proportion, as afore.* 3 Suppose a stone were 4 foot (or 48 inches) ouer, and 8 inches thick. Here 24 the halfe of 48, and 8, doe assigne 9 inches for the length desired. Therefore I say, that euery 9 inches in length of that stone, shall conteine 2 foot of solid measure. 4 Admit the stone were a yard square, that is, 36 inches broad, and 36 inches thick. Here both the dimensions, to wit, both breadth and thicknesse, are greater then any of those about the Trigon. Therefore I take 18 and 18, the halfe of each; and I finde them to meet in the space where you haue 5 inches and $\frac{1}{3}$. Therefore I say, that euery 5 inches, and $\frac{1}{3}$ part of an inch doth conteine 4 foot of stone.

To measure by that our Rular, being by the crossing of two lines giuen, to finde out the third; And the eye not able in many cases precisely to discerne at what parts of an inch that crossing is; some haue desired that these seuerall meetings might bee noted on the verge of the Rular, either aboue or beneath; but this being not to be done without confusion, I doe aduise them to haue recourse to this our Table, where they shall receiue satisfaction: And withall it shall teach them how to doe it. Other vses of this our Trigon, shall, God willing, shortly bee declared: for this straightnesse of roome will admit of no long discourse. Vale.

																							24	
																					23	3	24
																				22	3 141/529	3 3/23	23
																			21	3 69/121	3 105/253	3 3/11	22
																		20	3 45/49	3 57/77	3 93/161	3 3/7	21
																	19	4 8/25	4 4/35	3 51/55	3 87/115	3 3/5	20
																18	4 284/361	4 52/95	4 44/133	4 28/209	3 417/437	3 15/19	19
															17	5 1/3	5 1/19	4 4/5	4 4/7	4 4/11	4 4/23	4	18
														16	5 283/289	5 11/17	5 113/323	5 7/85	5 100/119	4 116/187	4 164/391	4 4/17	17
													15	6 3/4	6 6/17	6	5 27/38	5 2/5	5 1/7	4 10/11	4 16/23	4 1/2	16
												14	7 17/25	7 1/5	6 66/85	6 2/5	6 6/95	5 19/25	5 51/105	5 13/55	5 1/115	4 4/5	15
											13	8 40/49	8 8/35	7 5/7	7 31/119	6 6/7	6 66/133	6 6/35	5 43/49	5 47/77	5 59/161	5 1/7	14
										12	10 38/169	9 45/91	8 168/195	8 4/13	7 181/221	7 5/13	6 246/247	6 42/65	6 30/91	6 6/143	5 233/299	5 7/13	13
									11	12	11 1/13	10 2/7	9 3/5	9	8 8/17	8	7 33/57	7 1/5	6 6/7	6 6/11	6 6/23	6	12
								10	14 34/121	13 1/11	12 12/143	11 17/77	10 26/55	9 9/11	9 45/187	8 8/11	8 56/209	7 47/55	7 111/231	7 17/121	6 210/253	6 6/11	11
							9	17 7/25	15 39/55	14 2/5	13 19/65	12 12/35	11 13/25	10 4/5	10 14/85	9 3/5	9 9/95	8 16/25	8 8/35	7 47/55	7 59/115	7 1/5	10
						8	21 1/3	19 1/5	17 5/11	16	14 10/13	13 5/7	12 4/5	12	11 5/17	10 2/3	10 2/19	9 3/5	9 1/7	8 8/11	8 8/23	8	9
					7	27	24	21 3/5	19 7/11	18	16 8/13	15 3/7	14 2/5	13 1/2	12 12/17	12	11 7/19	10 4/5	10 2/7	9 9/11	9 9/23	9	8
				6	35 13/49	30 6/7	27 3/7	24 24/35	22 34/77	20 4/7	18 90/91	17 31/49	16 16/35	15 3/7	14 62/119	13 5/7	12 132/133	12 12/35	11 111/147	11 17/77	10 118/161	10 2/7	7
			5	48	41 1/7	36	32	28 4/5	26 2/11	24	22 2/13	20 4/7	19 1/5	18	16 16/17	16	15 9/57	14 2/5	13 5/7	13 1/11	12 12/23	12	6
		4	69 3/25	57 3/5	49 13/35	43 1/5	38 2/5	34 14/25	31 23/55	28 4/5	26 38/65	24 24/35	23 1/25	21 3/5	20 28/85	19 1/5	18 18/95	17 7/25	16 16/35	15 39/55	14 118/115	14 2/5	5
	3	108	86 2/5	72	61 5/7	54	48	43 1/5	39 3/11	36	33 3/13	30 6/7	28 4/5	27	25 7/17	24	22 14/19	21 3/5	20 4/7	19 7/11	18 18/23	18	4
2	192	144	115 1/5	96	82 2/7	72	64	57 3/5	52 4/11	48	44 4/13	41 1/7	38 2/5	36	33 15/17	32	30 6/19	28 4/5	27 3/7	26 2/11	25 1/23	24	3	
1	432	288	216	172 4/5	144	123 3/7	108	96	86 2/5	78 6/11	72	66 6/13	61 5/7	57 3/5	54	50 14/17	48	45 9/19	43 1/5	41 1/7	39 3/11	37 13/23	36	2
1728	864	576	432	345 3/5	288	246 6/7	216	192	172 4/5	157 1/11	144	132 12/13	123 3/7	115 1/5	108	101 11/17	96	90 18/19	86 2/5	82 2/7	78 6/11	75 3/23	72	1
1	2	3	4	5	6	7	8	9	10	11	12	13	14	15	16	17	18	19	20	21	22	23	24	

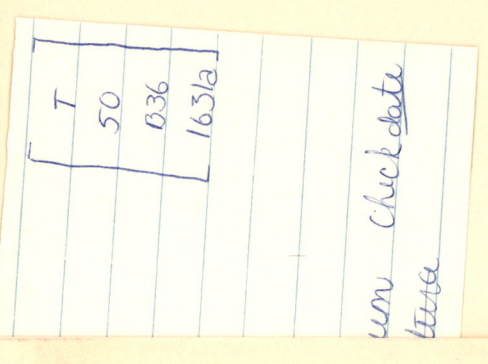